Janes Valdo Rodrigues Lima

Cost Approach to Austenitic Stainless Steel Milling

Janes Valdo Rodrigues Lima

# Cost Approach to Austenitic Stainless Steel Milling

## Projecting Manufacturing Costs

**Imprint**

Any brand names and product names mentioned in this book are subject to trademark, brand or patent protection and are trademarks or registered trademarks of their respective holders. The use of brand names, product names, common names, trade names, product descriptions etc. even without a particular marking in this work is in no way to be construed to mean that such names may be regarded as unrestricted in respect of trademark and brand protection legislation and could thus be used by anyone.

Cover image: www.ingimage.com

This book is a translation from the original published under ISBN 978-3-330-99797-4.

Publisher:
Sciencia Scripts
is a trademark of
Dodo Books Indian Ocean Ltd. and OmniScriptum S.R.L publishing group

120 High Road, East Finchley, London, N2 9ED, United Kingdom
Str. Armeneasca 28/1, office 1, Chisinau MD-2012, Republic of Moldova, Europe
Managing Directors: Ieva Konstantinova, Victoria Ursu
info@omniscriptum.com

Printed at: see last page
ISBN: 978-620-3-29827-7

**DEDICATORY**

To God. Every conquest is for Him and comes from Him.

This dedication is for my family: mum, sister. My support.

Families acquired here in São Luís who were present and important:

Family of Domingos Melo, Elza Barbara and Vitória, Pedro Osa, Joelson Neves.

**ACKNOWLEDGEMENTS**

There are so many people to thank, honestly, I hope I haven't forgotten anyone.

I want to thank the One who is the most important in my life, the First. The creator God, invisible but real. Lord of all things and holder of all wisdom and science. He is the first, Alpha and Omega. Without Him, these people mentioned below would not exist in my life and would not have the importance and love that they have. I thank you for everything, Lord Jesus. I love you with all my heart, soul and understanding.

To Professor Jean Robert Rodrigues, who accepted the challenge of being my supervisor and agreed to help me complete this course. He helped me with other projects, participation in congresses, etc.

To engineer Rômulo Barbosa, who co-guided me in this moment of decision and also guided the construction of this TCC. He gave up a few hours of sleep and time to help me develop it.

Right now I want to thank the woman who gave birth to life, the one who protected me as best she could. Who looked after me. She invested in me. She always thought of me. And she cheered for my successes in life. To my mum. To the one I love very much and will always love. Thank you for everything you've done.

To my family in general, my sister Jeane who is thrilled with my achievements and I with hers. To my grandfather Raimundo, who has been a partner in my success. And to all the other participants in this programme.

To my brothers and sisters in Christ, Joelson and Léia Neves, Domingos and Maria de Deus (Deuzinha), Elza and Vitória, who welcomed me into their homes and raised me as their own. To Manoel and Cleane who always cheer me on and ask about my academic training. To Mariane Souza, someone who motivated me a lot during this time, who was by my side and cheered me on during my achievements, and cried during my defeats and setbacks. God rewards you. Thank you for everything.

To my friends, university students and brothers in Christ. To Bruno Duarte, my partner since "Principe de Eng Mecanica" - the *Canada trip project*. *To* Maria Virlene, a friend of running, of calculations, of work that I forgot to do, where we shared frustrations, joy, hope and dreams. To my siblings acquired at UEMA, Jonatas Castro, Thaliane França and Clara Rayol, who were always there when we didn't see each other. And they will be friends for life, even after the doctorate and in heaven.

**SUMMARY**

Every day, companies are faced with challenges and are looking for the best options to determine how much lower the costs involved in manufacturing processes will be, machining materials, choosing the best product at the lowest cost, etc. Companies in all sectors are constantly looking for new technologies for their production chain in order to increase their manufacturing efficiency. This work consists of demonstrating the economic parameters in the Austenitic Stainless Steel Milling Process. As the quest for production efficiency is one of the main targets of any business, a much clearer and more explanatory analysis of the mechanisms that determine this ideal situation is necessary. A general analysis is also made of the characteristics of Austenitic Stainless Steel, with an emphasis on CN-3MN Superaustenitic Stainless Steel; showing specific properties, main applications, as well as economic advantages over other steels. In order to observe the economic conditions, various machining variables are used, some of which are totally empirical and others specific to the materials being machined. However, the most important unknown is the cutting speed, to which all the others relate, as it is present in most equations. In this way, this work highlights all these relationships between economic parameters and cutting speed. Milling serves as the 'background' for the relationships used.

Keywords: Austenitic stainless steel, Superaustenitic stainless steel, Milling, Costs, Cutting speed.

**SUMMARY**

# CHAPTER 1

INTRODUCTION

In series production, it is necessary to know which machining conditions will generate the lowest manufacturing costs, i.e. to determine the economic cutting conditions. An increase in cutting speed and feed rate will result in greater production, i.e. a reduction in the number of hours worked per piece, which implies a lower manufacturing cost. On the other hand, higher cutting speeds and feeds will increase the number of tools used per piece, as tool wear is increased [23].

For companies in the machining sector, the market has become increasingly competitive, forcing them to set sales prices for their services/products around those stipulated by the market versus the costs of HHT (Man Hours Worked), raw materials, cutting tools, cutting fluids, etc. used in production. As a result, companies in the machining industry need to survive by optimising their manufacturing processes, structuring their cost guidelines and sometimes even reducing their profit margins, while having to maintain the standard of quality that their products need to have by the standards of the competing market. This is a major dilemma that the metalworking industry has to deal with every day in order to build its production. Balancing an increase in cutting speed and feed rate with lower cutting tool replacement costs, while still taking into account production HHT [35].

One of the most important production processes for mechanical components is machining. Machining is a mechanical manufacturing process characterised by the removal of material, i.e. a raw material is subjected to the action of a tool that removes part of its material to adapt it to the desired geometry [15].

Milling is one of the most important operations in the mechanical machining process. It consists of removing excess material or overmaterial from the surface of a workpiece (chip removal) in order to build rectilinear flat surfaces or surfaces with a certain desired shape and finish. In milling, the excess material is removed from the workpiece by a combination of two movements carried out at the same time. One movement is the rotation of the tool around its axis, the other is the movement of the machine table, where the part to be machined is clamped. It is the movement of the machine tool table or feed movement that brings the workpiece to the cutting tool and makes the machining operation possible. The tool, called a milling cutter, has cutting edges (cutting edges) arranged symmetrically around an axis [4].

In history, research has first sought the best geometries for cutting operations. Then they looked for materials with better resistance characteristics and durability. Subsequently, materials were combined in new construction models in line with performance, cost and downtime reduction requirements in the production process. The fact that the milling cutter can take many different forms gives this operation versatility in terms of the geometries that can be generated. Many of the non-planar and non-revolving surfaces of parts and/or mechanical components can only be generated by milling [34].

Steel is characterised as the most versatile and important of metal alloys. In recent years, the production of stainless steel around the world has grown significantly. Sectors such as the food, mining, automotive and architecture industries are using stainless steel in numerous applications.

In addition to iron, carbon and chromium, austenitic steels have a high percentage of nickel in their chemical composition, whose function is to stabilise the austenite at room temperature. Another component that can replace nickel to stabilise austenite is manganese, but it must be added in greater quantities [7].

The development of new alloys for austenitic stainless steels is becoming increasingly important as the reduction of manufacturing costs and new applications are fundamental to maintaining the competitiveness of industries. A new alloy that does not require the addition of nickel to maintain a fully austenitic microstructure at room temperature is a very interesting proposal. The raw material used in steelmaking must be as cheap as possible, and Nickel appears to be the main factor responsible for the variation in price and cost of austenitic stainless steel, which makes cost and sales planning difficult for large companies [30].

With a predominantly austenitic structure, superaustenitic stainless steel shows better resistance to pitting, crevice and stress corrosion. This when compared to conventional austenitic stainless steels. What's more, its cost per kilo is less than half that of traditional corrosion-resistant alloys with the presence of nickel and it still offers excellent performance in a variety of highly corrosive environments [20].

Steel is produced in a wide variety of types and shapes, each of which efficiently fulfils one or more applications. This variety stems from the need to continually adapt the product to the demands of specific applications that emerge on the market, whether by controlling the chemical composition, guaranteeing specific properties or even the shape [29].

This work will be an approach to cost conditions and price decision-making that can be used to optimise economic considerations in machining, with milling as the mechanical process. The limits of maximum efficiency are determined, as well as the cutting speed for maximum profit. The theoretical basis presents concepts, models and forms for cost surveys currently in use in companies.

The goal of companies is to achieve a satisfactory level of profit on invested capital over the long term. In the case of large companies, the concept of satisfactory can be taken to mean maximum profitability on the capital invested over a given time horizon. In the case of small and medium-sized companies, satisfactory often means guaranteeing their survival over a much shorter time horizon [4].

The importance of being able to select the optimum manufacturing conditions has been recognised in the field of metal machining. The basic mathematical model that has been used in economic analyses is the unit cost model, or the analogous unit time model if costs are left aside. In association with these models, two criteria have been used to determine the optimum cutting conditions - one is minimum cost and the other is maximum production [38].

On the market, there are still a number of companies that are not yet concerned with the optimisation aspect of their processes; these are generally medium-sized and small companies. Most of them find it difficult to determine the ideal production conditions, due to a lack of knowledge or the difficulty of determining the fundamental parameters. The explanations found in the literature are almost entirely superficial, generally because they are not the focus of the study [5].

## 1.1. OBJECTIVES

### 1.1.1 General Objective

To analyse and find parameters for studying the economic aspect of milling Austenitic Stainless Steel, used in various mechanical applications.

### 1.1.2 Specific objectives

*F* Make connections between Austenitic Stainless Steel and the economic aspects of Machining;

*D* Determine the parameters of the economic aspect;

*C* Correlate the determinants of milling with production conditions;

*And* Finding the Conditions for Maximum Efficiency.

## 1.2. METHODOLOGY

The purpose of this work is to carry out an industrial analysis of the costs, operations and labour involved in manufacturing Austenitic Stainless Steel. It will evaluate the economic aspects that determine the final price of the finished product, the use of the best machining path in the milling process and the variations of the best result for the cutting speed. In order to realise the gain from the greater quantity produced and to realise the lower costs generated in the production process.

# CHAPTER 2

LITERATURE REVIEW

In this section, the theoretical background studied to initialise the work is presented. Many of the concepts presented here will be referenced throughout this work.

## 2.1 STAINLESS STEELS

### 2.1.1 Historical Context

The first steel works appeared practically at the same time as the industrial production of this material began, around 1780. There is a record of steel being used in the staircase of the Louvre museum in Paris, and a short time earlier, in 1757, a cast iron bridge was built in England. With advances in the steel manufacturing process around 1880, steel was already widely used in construction in the United States. In Brazil, the first steel structure was the bridge over the Paraíba do Sul river, in the state of Rio de Janeiro, in 1857 [36].

At the beginning of the 20th century, metallurgists noticed that chromium was more attractive to oxygen than iron and decided to add the element chromium to steel. Studies have shown that when at least 10% chromium is added to steel, this element unites with oxygen to form a thin, continuous and transparent layer on the surface of the steel, which prevents further oxidation. This transparent layer reappears when the surface suffers any damage, such as scratches, wear or denting [2]. This high resistance to chemical attack, common to all stainless steels, is a property that is also called passivity - which consists of the formation of a continuous, thin and impermeable layer of chromic oxide that protects the underlying material from corrosive attack [18][1].

The man responsible for developing stainless steel, according to the story, was Englishman Harry Brealy in 1912. He observed some metallographic characteristics after carrying out experiments with an iron-chromium alloy and realised that the alloy was more resistant to the reagents commonly used in metallography. He labelled the alloy "Stainless Steel", because the steel was not 'stained' when subjected to metallographic attacks [1][25][5].

In Germany that same year, Eduard Maurer claimed that an alloy, also made of iron-chromium, developed by Brenno Straus, was resistant to aggressive vapours; vapours from his laboratory [1][25][5].

The advances made since then in mechanical manufacturing processes and the refining of metal alloys have led to the development of stainless steels with different chemical compositions, microstructures and chemical and mechanical properties that best suit their future applications. These steels are divided into different classes that vary according to the chemical elements present in them; these elements are responsible for stabilising the ferritic or austenitic microstructure or both [33].

Stainless steels are iron-based alloys that contain a minimum of approximately 11% chromium as the main alloying element [3]. They are steels that do not oxidise in normal environments. Some stainless steels have more than 30 per cent chromium or less than 50 per cent iron. Some other elements such as Nickel (Ni), Molybdenum (Mo), Copper (Cu), Titanium (Ti), Aluminium (Al), Silicon (Si), Niobium (Nb), Nitrogen (N) and Selenium (Se) can be added to obtain particular mechanical characteristics [18].

The chemical industry and high-temperature applications then had a new class of materials suitable for their installations in aggressive environments. The research that was carried out clearly indicated the impact that these materials represented. In 1934, 56,000 tonnes were produced and in 1953 world production exceeded one million tonnes. Between 1950 and 1980, the production of stainless steels increased 20-fold; around two thirds of this production was authentic stainless steels [27][16].


2.1.2 Corrosion resistance and passivity.


Stainless steels are alloys of iron (Fe), carbon (C) and chromium (Cr) with a minimum of 10.5% Cr, according to CARBÓ, and 11% according to other authors. Other metallic elements are also part of these alloys, but Cr is considered the most important element because it gives stainless steels high resistance to corrosion. In rural atmospheres, see table 2.1, with low levels of contamination, there is a significant decrease in the oxidation rate of these alloys as the amount of Cr present in them increases, see figure 2.1. With 10.5% Cr, the alloy does not suffer atmospheric corrosion under these conditions and this is the criterion used to support the definition given at the beginning of this text for stainless steels [6].

Table 2.1: Influence Factors for Atmospheric Corrosion [19].

| Atmosphere | Relative corrosion |
|---|---|
| Dry countryside | 1 - 9 |
| Navy | 38 |
| Industrial (marine) | 50 |
| Industrial | 65 |
| Industrial, heavily polluted | 100 |

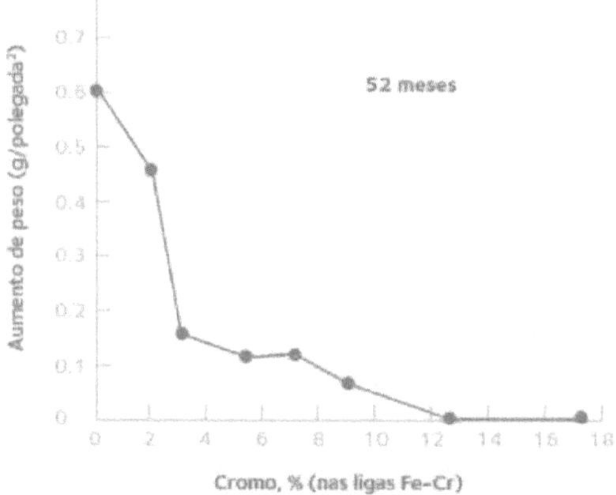


Figura 2.1: Effect of the presence of chromium in stainless steels on the oxidation rate [6].

Corrosion is defined as the deterioration of a material, usually metallic, by the chemical or electrochemical action of the environment, whether or not combined with mechanical stress [19]. In general, all metals (with a few exceptions) have a great tendency to react in the presence of the environment, forming oxides, hydroxides and other chemical compounds [6]. Corrosive processes are electrochemical and present identical mechanisms to each other, as they always consist of anodic and cathodic areas, in which an electron current and a current of ions circulate, causing a loss of mass and a mode of attack on the material that can take place in different ways or environments [37].

Corrosion can be uniform or localised. Localised corrosion is more difficult to predict and control, as there are several types of localised corrosion, such as pitting, galvanic corrosion, hydrogen-induced cracking, fatigue corrosion, intergranular corrosion and stress corrosion [37].

Uniform corrosion is the attack of the entire metal surface that is in contact with the corrosive medium, leading to a reduction in thickness. This form of corrosion generally occurs due to locally-acting micropiles and is the most common, occurring mainly in structures exposed to the atmosphere [19].

Uniform corrosion is the easiest corrosion to verify, especially when it comes to internal corrosion in equipment or installations, as the reduction in thickness is approximately the same across the entire metal surface. On the other hand, localised corrosion is corrosion that occurs in specific areas and sometimes in places that are difficult to detect, which can result in cracks [37].

We should emphasise that nature permanently transforms metals into their compounds through spontaneous reactions in which energy is released. This is why we find metals in nature in the form of oxides, hydroxides and salts of these metals. The steel industry has the opposite mission:

to transform these ores into more or less pure metals or their alloys. The reactions in the steel industry are the opposite of those that occur in nature and, for this reason, they are not spontaneous and require energy to be carried out, see figure 2.2 [26].

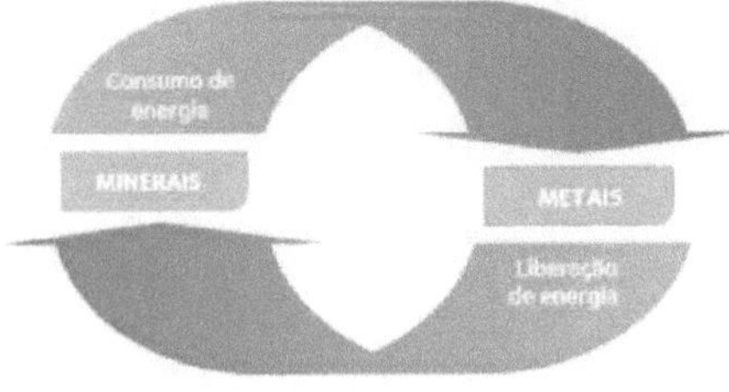


Figura 2.2: Process of transforming Raw Materials *into Metals and Metals into Raw Materials* [6].

Naturally, these metals and alloys obtained in the steel industry tend, over time, to naturally transform into compounds of the same, and this process is known as corrosion [6].

Due to the cost of corrosion, which in some countries is considered to be around 3% of GDP [6]. Corrosion problems are frequent and occur in a wide variety of activities, such as in the chemical, oil, shipbuilding, construction, automotive, air, rail, metro, sea and road transport and communication industries, dentistry, medicine and works of art [19].

The economic losses that affect these activities can be categorised as direct (a) the cost of replacing parts or equipment, (b) the cost of maintaining protection processes and indirect (a) accidental stoppages, (b) loss of efficiency, (c) product contamination, (d) product loss and (e) project oversizing. [19].

The phenomenon of passivity has been studied for many years and there have been (and still are) various interpretations of it. It is known that the formation of these films is favoured by the presence of oxidising media. The passive film of stainless steels is very thin and adherent. Films formed in oxidising media (such as nitric acid, often used in pickling baths) are more resistant. Stainless steels form and maintain passive films in a wide variety of media, which explains the high corrosion resistance of these materials and the large number of alternatives that exist for their use [6].

In general, stainless steels have good resistance to corrosion in oxidising media (which facilitate the formation and preservation of passive films). The corrosion resistance of these materials is poor in reducing media (which do not allow these films to form or destroy them). The fact that a large number of "aggressive" media act in the field of passivity explains the high corrosion resistance of stainless steels and the wide possibilities for their use in various applications [6].

Other elements besides chromium can be added to steel to increase its resistance to corrosion in specific media. However, chromium is the main element responsible for the corrosion resistance of these steels and also for the formation of the protective layer on the surface, which gives rise to the phenomenon known as passivation. The presence of this layer begins to be noticeable at chromium contents of 10% or more and as this content increases, so does the stability of the layer. For certain alloys, the chromium content can reach 29%. Chromium has a marked influence on mechanical properties and these properties are improved with the presence of just 2% chromium in the alloy. However, when the chromium content in the alloy exceeds 29%, these properties may be compromised [12].

The increasing presence of this element reduces the alloy's ability to be hardened by tempering, as it makes the austenite stability region smaller and smaller, and consequently increases the ferrite stability region (alphagen element). This can be verified by studying the effect of chromium on the austenite zone of the Fe-C diagram, shown in Figure 2.3 [26]. [9].

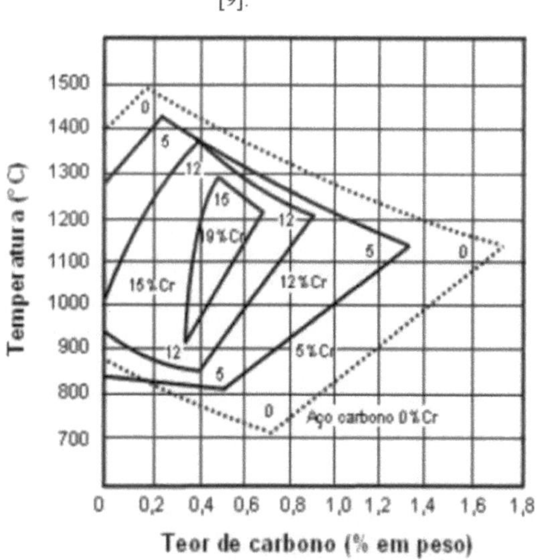


Figure 2.3: Effect of chromium content on the austenitic field of the Fe-C system

The simplest and most widely used classification of stainless steels is based on their microstructure, chemical composition and crystallographic factors, which are related to their mechanical and chemical properties. These steels are divided into five families based on their general characteristics in terms of mechanical properties and resistance to corrosion [12]:

*A* Austenitic (AISI 200 and 300 series);

Femtics (AISI 400 series);

*F* Femtico - austenitic *(Duplex)*

*M* Martensitic (AISI 400 series);

*E* Precipitation hardening - PH.


### 2.1.3    Ferritic and Martensitic Stainless Steels


In stainless steels, two elements stand out: chromium, which is always present because of its important role in corrosion resistance, and nickel, because of its contribution to improving mechanical properties [6].

Although there are different classifications, some of which are more complete than the one presented here, we can, in principle, divide stainless steels into two main groups: **the 400 series** and **the 300 series**. The 400 series are ferritic stainless steels, magnetic steels with a CCC structure, basically Fe-Cr alloys. The 300 series are Austenitic Stainless Steels, non-magnetic steels with a CFC structure, basically Fe-Cr-Ni alloys [6]

All stainless steels also contain carbon and other elements found in all steels, such as silicon (Si), manganese (Mn), phosphorus (P) and sulphur (S)[6].

Stainless steels in the 400 series can be divided into two sub-groups: ferritic steels, which generally have higher chromium and lower carbon, and martensitic steels, in which lower chromium and higher carbon predominate (compared to ferritic steels) [6].

In martensitic stainless steels, the carbon is in a certain concentration that allows the transformation of ferrite into austenite at high temperatures. They are characterised by being chromium steels, between 11.5% and 18%. During cooling, austenite is transformed into martensite. Martensite is a carbon-rich, brittle and very hard phase. These steels are manufactured and sold by the steel industry in the annealed state, with a ferritic structure, low hardness and good ductility. Only after a tempering heat treatment will they have a martensitic structure, being very hard and not very ductile. But in these conditions (tempered) they will be resistant to corrosion [6,9].

Within this group, three further classes can be considered [9]:

*a) low carbon, also called the "turbine" type b) medium carbon, also called the "cutlery" type c) high carbon, also called the "wear-resistant" type.*


Among martensitic stainless steels, the best known is 420, with just over 12% Cr and approximately 0.35% C [6].

Other elements have also been added to improve the characteristics of martensitic materials, such as nickel and [9]:

- titanium, which reduces the tendency to grain growth and increases weldability; niobium acts in the same way;

- molybdenum, which, at 1 to 2 per cent, significantly increases resistance to the action of dilute acids, organic acids, etc;

- aluminium, which apparently reduces grain growth at high temperatures.

Ferritic stainless steels, figure 2.4, are iron-chromium alloys containing 10.5 to 30 per cent chromium with a low carbon content. They are magnetic and have good resistance to corrosion in less aggressive media, in media containing chlorides and excellent resistance to stress corrosion, which are major advantages over some austenitic stainless steels, good ductility and reasonable weldability. These steels have a microstructure made up of ferrite, a solid solution of carbon in iron, with a Cubic Centred Body (CCC) crystal structure. In addition to the iron and chromium present in these alloys, there are other elements such as nickel, molybdenum, titanium and others, which increase the corrosion resistance and weldability of these alloys. Chromium is the element responsible for the stability of ferrite, which increases as the content of this element increases. [12].

Type 430 is the most widely used [6,9], due to its great resistance to the action of acids, especially nitric and organic acids, and to the action of seawater. This is the only type of steel in the ferritic group that is not entirely ferritic and can be slightly hardened by rapid cooling [9].

In addition, due to their low nickel content, these steels have a low cost, making them extremely attractive for some industrial applications. On the other hand, a disadvantage in the application of ferritic stainless steels has been their loss of corrosion resistance, as well as ductility and toughness when exposed to a certain temperature, mainly due to the formation of phases which will be illustrated later [12].

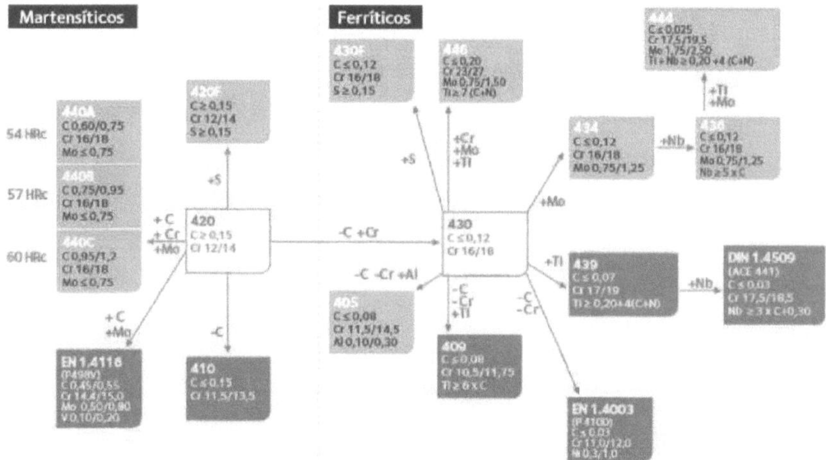


Figure 2.4: 400 Series stainless steels [6].

### 2.1.4 Parameters that influence stainless steels

According to Chiaverini, the simplest and most widely used classification of stainless steels is based on their microstructure at room temperature. Under these conditions, we can highlight hardenable steels (Martensitic) and non-hardenable steels (Ferritic and Austenitic) [9].

The main composition of stainless steel and its main characteristics are well known. However, adding other elements to this alloy adds more specific characteristics to it. Elements such as Nickel, Molybdenum, Niobium and Titanium [5]:

Nickel: modifies the structure of the material, which has better ductility, mechanical and hot resistance, and weldability; Chromium and nickel are the main elements in stainless steels;

*M* Molybdenum and Copper: increases resistance to wet corrosion;

*S* Silicon and Aluminium: improve resistance to oxidation at high temperatures;

*T* Titanium and Niobium: stabilising elements in Austenitic Steels, preventing Chromium depletion through precipitation in the form of carbides during heat treatment;

*O* Others: to modify more specific characteristics of stainless steel, elements such as manganese, nitrogen, cobalt, boron and rare earths are used [5].

Alloying elements are added to stainless steels for specific purposes. There is an important factor not traditionally mentioned, which is the effect of alloying elements on Nitrogen solubility. The Iron Valence Layer combined with that of each element dictates which elements are favourable (increasing solubility) to Nitrogen. Electron-poor elements such as chromium, manganese and vanadium increase solubility, while elements with excess electrons decrease it [30].

Note that in figure 2.5, as the chromium content increases, the austenitic range decreases, until it practically disappears at around 20% chromium. This leads to the conclusion that, as the chromium content increases, the composition of Fe-C-Cr alloys that will allow total hardening to be achieved is reduced to increasingly narrow limits [9].

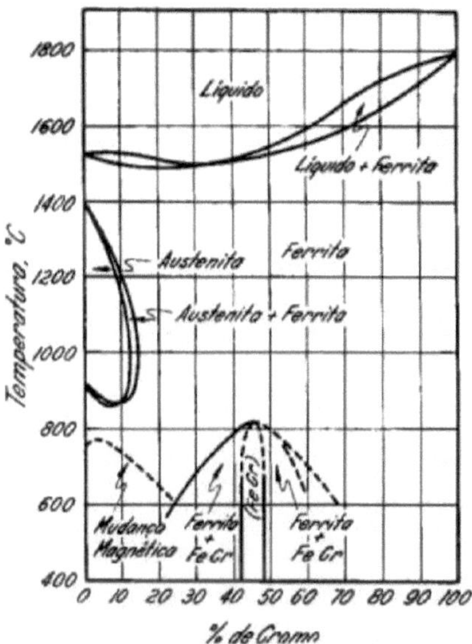

Figura 2.5: Fe-Cr alloy constitution diagram [9].

Nickel, on the other hand, has the opposite effect to chromium (a gamma-genetic element), because as its content increases, the austenite stability zone increases and, consequently, the range of existence of this phase increases, extending to room temperature. Thus, when both elements are present, a compromise situation results and the two allotropic forms, austenite and ferrite, can develop at their appropriate temperatures [12], as shown in figure 2.6.

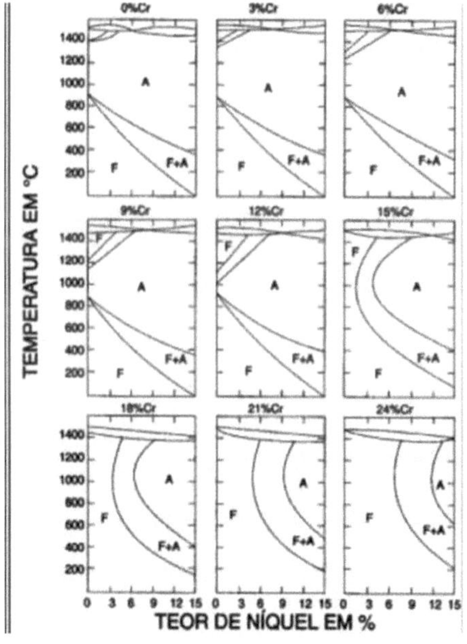

Manganese was originally introduced as a substitute for Nickel due to its supply problems or price instability. To stabilise the Austenite, the Nickel content can be reduced to approximately 4% by weight by adding 2-6% Mn. However, with this substitution, the steels do not perform as well in aggressive environments [30].

When manganese is added in small quantities, associated with the presence of nickel, it substantially improves the functions attributed to nickel, while molybdenum combined with chromium has a great effect on the stability of the passivation film in the presence of chlorides, as can be seen in Figure 2.7: Effect of Mo and Ni on the austenitic field [12].

Chemical elements such as molybdenum, vanadium, tungsten, silicon, niobium and aluminium, which are alphagen elements, promote the formation of the ferritic phase in stainless steels. On the other hand, elements such as nickel, manganese, carbon, nitrogen, copper and cobalt, which are gammagen elements, promote the formation of the austenitic phase [12].

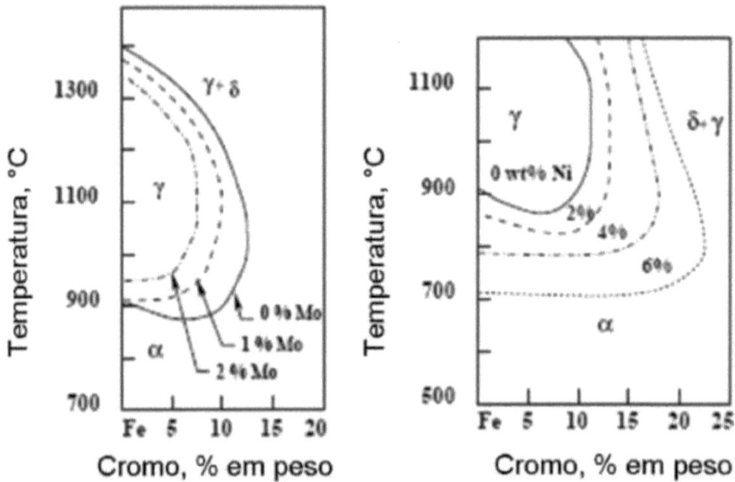


Figura 2.7:  Segments of the Fe-Cr diagram showing the effect of Mo and Ni on the austenitic field [12].

The role of carbon is directly related to the type of steel (martensitic, austenitic and ferritic). In austenitic steels, carbon favours the formation of austenite, but is harmful in terms of sensitisation and the occurrence of intergranular corrosion. In addition to these elements, stabilising elements such as titanium, niobium and tantalum can also be added, which have a great affinity for carbon. These elements are used to prevent or hinder sensitisation due to the formation of chromium carbides. Therefore, the carbon in the alloy combines with these elements to form titanium (TiC), niobium

(NbC) or tantalum (TaC) carbides [12].

Always a cause for concern in austenitic stainless steels, carbon acts by increasing mechanical strength via solid solution in non-austenitic grades and by increasing strength via precipitation when Nb, Ti or V are present (PH steels). In austenitic steels, it must be very well controlled to avoid sensitising the material; however, it inhibits the formation of Martensite induced by plastic deformation [30].

Titanium, niobium and vanadium are elements present in so-called stabilised steels, as they have a great affinity for carbon, preventing sensitisation. In addition, they have a great effect on increasing creep resistance in austenitic steels due to the precipitation of refined carbides in the intergranular region; however, they reduce ductility when already subjected to creep. The additions of these elements must be in accordance with the percentage of carbon, and with great care when nitrogen is present as it also tends to form nitrides [30].

2.1.5 Austenitic stainless steel

Austenitic stainless steels form the largest group of stainless steels in use, accounting for around 65 to 70 per cent of the total produced [22], and of the stainless steel grades, it is the one that is most resistant to oxidation. This class, along with ferritic, cannot be hardened by heat treatment [13].

On 17 October 1912, Krupp, headed by Brenno Straus, filed patent application DRP 304126 with the patent office of the German Empire in Berlin: "Manufacture of objects requiring high resistance to corrosion". This patent was shortly afterwards, on 20 December 1912, supplemented by another: DRP 304159. This gave birth to V2A austenitic stainless steels (V for Versuch, meaning experience, A for Austenit) containing 20% chromium, 7% nickel and 0.25% carbon. Krupp was the first company to commercialise stainless steels. In 1914, the firm had already supplied 18 tonnes of V2A to an aniline and soda factory; stainless steels were immediately adopted in the country's synthetic ammonia factories. The alloys ferrochrome (17% chromium) and ferrochrome-nickel (18% chromium and 8% nickel) were widely used in the 1920s and 1930s in the United States, England and Germany, in ammonia and nitric acid factories [25,16].

Austenitic stainless steels are very useful materials in industry and are very well used in process equipment, due to their excellent chemical properties such as high resistance to corrosion and resistance to high temperatures. However, they are difficult materials to machine and austenitic stainless steels are in the class of the most difficult to machine [28], and therefore cost more to process. Due to the cost of corrosion, which in some countries is considered to be around 3% of GDP, people have been working for a long time to reduce costs by creating barriers against corrosion to at least minimise these problems, since it is impossible to eliminate them [6].

This resistance to corrosion, combined with ductility, good weldability and good performance over a wide temperature range are some of the reasons for using austenitic stainless steel. With a chromium content of over 11%, the steel has a very adherent surface oxide layer, providing a passive layer. However, the formation kinetics of this oxide layer, its stability and its ability to form again after suffering external damage, are closely linked to the addition of other alloying elements such as Nitrogen, Molybdenum and Nickel itself. This material is widely used in cryogenic applications due to its lack of a ductile-fragile transition and the drop in toughness as the temperature decreases, making it a material with a wide range of potential uses [30].

Austenitic stainless steels are included in the group of steels with the best mechanical resistance properties at high temperatures, as the coefficient of thermal expansion is approximately 60% higher and the thermal conductivity is approximately 30% higher when compared to ferritic stainless steels. Its CFC structure has a major influence, because it favours the material's excellent impact resistance values, as well as favouring the non-occurrence of the ductile-fragile transition phenomenon [20].

The machinability of a material is understood as the ease or difficulty of removing a material during the process and can be assessed through machining force (Fu), cutting temperature (Tc), surface finish, chip control, cutting tool wear rate [10] and other analyses, as seen in section **2.2** of this paper, depending on the need, availability of infrastructure, facility. We can also take into account the mechanical vibration (Vb) of the workpiece-tool system, acoustic emission signals, among others [16].

Austenitic stainless steels are iron-chromium-nickel ternary alloys with chromium contents of between 16 and 26 per cent and nickel contents of between 7 and 22 per cent. They have good mechanical properties, good weldability, cold workability and corrosion resistance. These steels can also be hardened by cold deformation and, in this state, the most common types become magnetic. The addition of alloying elements such as molybdenum and the reduction in carbon content improve their resistance to corrosion, especially pitting corrosion [28]. Nickel is a strong austenite former in these steels, when added in small quantities it improves the toughness and weldability of the alloy and accelerates the formation of the protective chromium oxide layer [19, 12].

Austenitic stainless steels, which are non-magnetic steels with a face-centred cubic (CFC) structure, generally consist basically of Fe-Cr-Ni. Mo (Molybdenum) and Mn (Manganese) are added depending on the application of the end product and the environment to which the material will be exposed. It should be noted that all stainless steels always contain carbon and other elements such as silicon (Si), manganese (Mn), phosphorus (P) and sulphur (S) in specific quantities according to their classification [6].

Austenitic stainless steels show complex microstructural changes when they undergo plastic deformation, as well as major variations in mechanical strength and corrosion resistance. These changes are due to the martensitic transformation induced by deformation and the low stacking fault energy (EFE) presented by this class. The low EFE is also a determining factor in the formation of maclas in austenitic steels, which also serve as nucleation sites for martensitic transformation [30].

This change is extremely sensitive to chemical composition, temperature, stress state and grain size. Normally, CFC metals show little influence of temperature on yield strength; however, the rate of hardening is greatly affected by the stacking fault energy, which is greatly influenced by temperature [30].

Figure 2.8 shows some variations of austenitic stainless steels, based on AISI 304, where elements are added or removed for specific purposes, where the use of N and Mn in the 200 series are already applied to increase resistance; however, Ni is still present (approximately 5%) [30]. Table 2.2 shows the variation in the alloying elements of austenitic stainless steels, and figure 2.8 shows this variation.

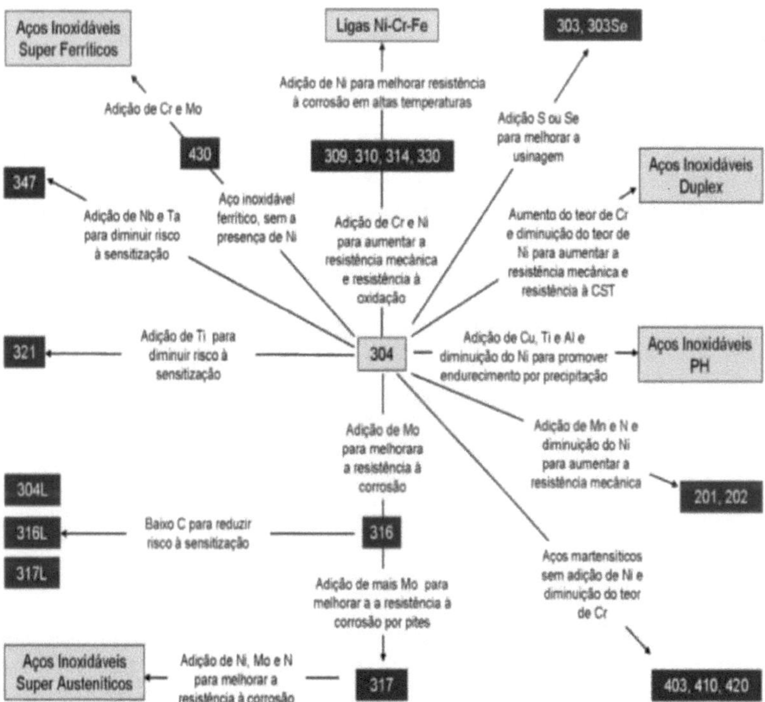


Figura 2.8: Diagram illustrating alloys and how the addition of elements modifies their properties [30].

Table 2.2: Variation in the composition of austenitic stainless steel alloys [16].

21

| Cr | Ni | C | P | Mn | S | Yes | Others |
|---|---|---|---|---|---|---|---|
| 16-30% | 6 - 22% | <0,03% | <0,06% | 2% - 300 Series<br>5-16% - 200 Series | <0,03% | <2% | <0,4% N<br>< 0.4%Mo<br>10*C>Nb+Ta |

### 1.1.6 Austenitic stainless steel with the presence of nickel

The development of new stainless steel alloys is the focus of attention for the world's major steel suppliers. Traditionally, stainless steels are presented in five different grades, as seen in section 2.1.2 of this paper:

austenitic, ferritic, duplex, martensitic and precipitation-hardened. Each class has specific characteristics for use in certain applications and, among these, the austenitic class has the largest share of the stainless steel market. It also has the highest production cost due to the raw material and its processing [30].

These reasons have led to new alternatives being developed to reduce production costs. The most costly input in this material is the addition of nickel, which accounts for between 30 and 40 per cent of the total cost of raw materials, an element that has been highlighted for its instability of supply and price, limiting long-term sales. Nickel is the traditional and main stabilising agent for Austenite, in the 300 class, and is normally added at levels above 8% to maintain the austenitic microstructure at room temperature [30].

Environmental and health issues are also frequently debated with the aim of changing the use of nickel, from pollution during extraction and processing to allergies when in contact with human skin. The sum of these reasons has led to the search for new options for stabilising Austenite without the addition of nickel. In the use of the 200 series, when the addition of nickel has been reduced and, for the first time, they become an effective set, both in stabilising austenite and in increasing resistance by solid solution and hardening; however, without martensitic transformation which, in some situations, induces the originally non-magnetic material into a magnetic state, changing its original characteristics and consequently changing its performance in service [30].

### *1.1.7 Austenitic stainless steel without the addition of nickel*

In recent years, the development of stainless steels without the addition of nickel has attracted a lot of attention. Reducing the use of nickel has already been widely used in the 200 series, where the addition of nitrogen and manganese to stabilise the austenite has made it possible to reduce the Ni content from 8% to between 3% and 6%. However, this grade did not match the performance of the most widely used austenitic stainless steel, AISI 304, when used in environments that required greater resistance to corrosion and in biomedical applications [30].

Used in structural applications, this series of steels, the 200 series, are the result of partially replacing nickel with manganese. The corrosion resistance of these alloys (Fe-Cr-Ni-Mn) is lower than that of equivalent 300 series steels. Some more recently manufactured 200 series alloys containing copper (Cu) allow some of these steels to be used in other applications (not just structural), such as deep drawing [6].

One of the reasons for the focus on reducing or even replacing elements such as nickel in austenitic alloys is that the presence of nickel increases the cost of obtaining the raw material and also makes austenitic steel difficult to machine, causing a loss of productivity, according to [14]:

fHeavy tool wear with poor surface finish quality;

*p* poor chip output characteristics (breakage), causing congestion in machining centres;

*b* low cutting speeds leading to low productivity.


Nickel is used to stabilise the austenitic band in stainless steel. However, due to its high cost and fluctuating price, steelmakers have been forced to invest in alternative production methods or elements for this purpose, as shown in figure 2.9. This variation often jeopardises the conclusion of long-term contracts, making it difficult to plan production, logistics, sales, etc [30].

In other sectors, such as biomedicine, the choice of austenitic steels without the addition of nickel is also quite intense. This is because some people suffer from skin allergies when they come into contact with this type of nickel-containing material. This makes nickel-free steel highly preferable [30]. This demand has fuelled the development of austenitic stainless steels and even biocompatible non-ferrous materials.

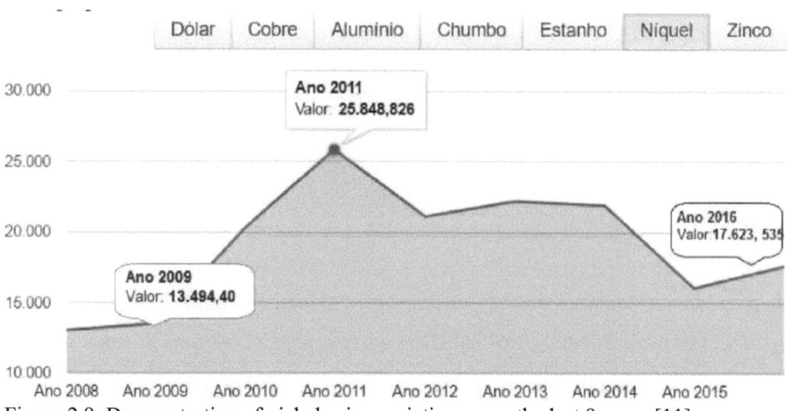


Figure 2.9: Demonstration of nickel price variations over the last 9 years [11].

According to Hanninem et al in [30], nitrogen in solid solution is the most beneficial element for increasing strength without affecting ductility and toughness, as long as the solubility limit is not exceeded, in commercial alloys usually less than 0.9%; however, this value depends on the chemical composition. Once the limit is exceeded, nitrides can form or pores can form.

Corrosion resistance is notoriously the most talked about issue in austenitic stainless steels and the addition of nitrogen is seen as the most promising alternative for improving performance in aggressive environments, mainly due to its repassivating or stabilising effect on the passive layer . Studies have shown that alloys with a combination of N and Mo show excellent results against pitting corrosion, both in the emergence of pitting and in its growth, where N appears to be an important repassivating agent for newly initiated pitting and Mo acts as a suppressor of initiation [30].

This new alloy, with the presence of nitrogen to replace nickel, is becoming commonplace in the industry and its introduction is not trivial. The solubility of nitrogen in liquid steel is low and the manufacturing processes used involve high pressures and require specially designed, expensive equipment.

A new family of steels with no added nickel, featuring variations of the combination of manganese and nitrogen as austenite stabilising elements. Nitrogen is a strong hardener of austenite by interstitial solid solution and, together with Mn, intensely lowers the stacking fault energy of austenite, as well as stabilising this phase, hindering the formation of martensite induced by plastic deformation and therefore increasing the rate of hardening [30].

Currently, a balance has to be made when nitrogen is added concurrently, since Mn greatly increases the solubility of N in austenite. Nitrogen is one of the strongest stabilisers of austenite, acting together with carbon to prevent induced martesite. It also acts to increase the creep life of austenitic steels. It can act like carbon in stabilised steels that precipitate Ti and Nb carbides, but with much greater strength-enhancing power than carbon. Its effect is beneficial in the passive layer, but its mechanism of action has not yet been identified [30].

As seen in section 2.1.4, titanium, niobium and vanadium are elements present in so-called stabilising steels. And their additions must comply with the percentage of carbon, and even more carefully when nitrogen is present, as it can form nitrides that are harmful to the composition.

### *1.1.8  Superaustenitics - CN-3MN in the **spotlight***

With a predominantly austenitic structure, superaustenitic stainless steel shows better resistance to pitting, crevice and stress corrosion. This when compared to conventional austenitic stainless steels. What's more, its cost per kilo is less than half that of traditional corrosion-resistant alloys with the presence of nickel, as shown in Table 2.3, and it also offers excellent performance in a variety of highly corrosive environments [20].

Known as *superaustenitic 6% Mo*, the alloys that belong to this group were first developed to resist localised corrosion in marine temperatures and environments. But they have other areas of use, such as: aerospace, biopharmaceuticals, beverages, nuclear, oil and gas production, food processing, heat treatment furnaces, pharmaceuticals and paper [20].

The 6% Mo superaustenitic contains relatively high levels of nitrogen, chromium and molybdenum and has a higher level of strength and ductility when compared to conventional austenitic stainless steels, especially the 300 series [20].

Table 2.3: Comparison of the average cost of a level alloy versus superaustenitic [20].

| Material | League | Cost/Kg(R$) |
|---|---|---|
| Nickel alloy | ASTM A494 Grade CW-6MC | 112 |
| Superaustenitic | ASTM A351 Grade CN-3MN | 55 |

These steels are also considered high performance austenitic steels. The ASTM A744 CN3MN alloy (UNS J94651), which belongs to this superaustenitic group, is one of the material options used in the manufacture of pump components operating in the oil sector. It is an Fe-Cr-Ni-Mo alloy [20].

During the manufacturing process, care must be taken to avoid the formation of certain types of precipitates and phases and for the steel to achieve the desired mechanical properties and resistance to corrosion. Although the addition of nitrogen helps to slow down the formation of phases other than austenite, their formation is inevitable, especially in the casting process of thick parts, where cooling is slower. It is known that at temperatures between 650 and 980°C, various precipitates appear in the austenite structure [31].

The physical and mechanical properties of CN3MN superaustenitic steel at room temperature are described in Table 2.4 and compared to the values of conventional cast austenitic stainless steel, according to ASTM A351 CF8M (type 316).

Table 2.4: Mechanical and physical properties of CN3MN and CF8M steels [20].

| Properties | Austenitic CF8M | Superaustenitic CN3MN |
|---|---|---|
| Density (g/cm$^3$) | 7,7 | 8,1 |
| Melting point (approx) (°C) | 1400 | 1320-1390 |
| Modulus of elasticity (E=MPa) | 195 | 195 |
| Poisson's ratio | 0,29 | 0,29 |
| Thermal conductivity (W/mk) | 16 | 12 |
| Electrical resistivity (pQ.m) | 0,82 | 0,89 |
| Minimum tensile strength (MPa) | 485 | 550 |
| Minimum yield strength (MPa) | 205 | 260 |
| Minimum elongation (%) | 30 | 35 |

Table 2.5 shows the chemical composition of this steel, which can be compared with the chemical composition of austenitic steels in Table 2.2.

Table 2.5: Chemical composition (%) of CN3MN superaustenitic stainless steel [20].

| Cr | Ni | C | P | Mn | S | Yes | Ass | N | Mo |
|---|---|---|---|---|---|---|---|---|---|

| 20 a 22 | 23,50 a 25,50 | <0,03 | <0,04 | <2,00 | <0,01 | <1,00 | <0,75 | 0,18 a 0,26 | 6 a 7 |
|---|---|---|---|---|---|---|---|---|---|

Finally, we can check the pitting resistance property, PREN, of CN3MN superaustenitic steel with other stainless steels in Table 2.6. The PREN value is obtained using the percentage of the elements chromium, molybdenum and nitrogen.

Table 2.6: Resistance to pitting corrosion for some steels [20].

| Name | ASTM standard | Cr (%) | Mo (%) | N(%) | $PRE_N$ |
|---|---|---|---|---|---|
| Austenitic | Gr.CF-8M | 19,0 | 2,2 | 0,08 | 27,5 |
| Duplex | Gr.3A | 25,5 | 2,1 | 0,20 | 35,6 |
| Superaustenitic | Gr.CN-3 MN | 21,0 | 6,5 | 0,22 | 46,0 |
| SuperDuplex | Gr.5A | 25,0 | 4,5 | 0,20 | 43,0 |

## 2.2 MACHINABILITY OF STAINLESS STEELS

The machinability of stainless steels is generally influenced by the material's alloying elements, the heat treatment carried out and the material's manufacturing process (forging, casting, etc.). To assess machinability there are also other important criteria, such as the metallurgical state of the part, hardness, chemical composition, thermal conductivity, mechanical properties and possible hardening. It is estimated that 15% of all mechanical components manufactured worldwide are derived from machining processes, while milling machining in the mould and die sector is one of the most widely used techniques [32].

Stainless steels are known for their difficult machinability, which translates into machining with a short tool life, limited material removal rate (Q), high cutting forces, high power consumption (due to their mechanical strength at high temperatures), rapid hardening during machining and reactivity with most tool materials when machining at high cutting speeds. The presence of elements such as chromium, nickel and molybdenum causes high plastic deformation, promoting high cutting and feed forces [22].

Because stainless steel comes in various crystalline structures, the machining characteristics vary for each type of steel. There are also other factors that can be described as increasing the difficulty of machining stainless steels, according to Krabbe [22]:

*B* Low thermal conductivity;

High tenacity;

*P* Presence of abrasive carbide particles in alloyed stainless steels, contributing to tool wear;

*T* Tendency to the formation of post-cutting edges (PCA), which unlike conventional steels are present at high cutting speeds due to high fracture toughness, high ductility and rapid hardening.

We can see the machinability characteristics for each existing type of stainless steel, as described in Table 2.7.

Table 2.7: Machinability characteristics of the various types of stainless steels [22].

| CLASS | GENERAL CHARACTERISTICS AND MACHINABILITY |
|---|---|
| **Ferritic (ABNT 430, 430F, 431, 444)** | Better machinability for alloys with low chromium content. Very thin and highly deformed chips for alloys with a higher Cr content. |
| **Martensitic (ABNT 403, 410, 416, 420F, 440)** | Better machinability for quenched alloys with low carbon content. Strong influence of hardness level and nickel and carbon content. |
| **Austenitic (ABNT 303, 304L, 310, 316, 316L)** | High tendency to form a false cutting edge. Difficult chip removal (thin and adherent). Hardened cutting surfaces. |
| **Precipitation hardening (UNS - S13800, S15500, S17400).** | Machinability varies for each type of alloy and for each level of material hardness achieved through precipitation hardening treatment. Machinability limited by relatively high hardness, improved with tempering treatment. |
| **Duplex (Ferritic + Austenitic - ABNT 318, 329, 325).** | Machinability limited by the level of mechanical resistance. High level of hardness. Few alternatives for versions with improved machinability. |

Austenitic stainless steels are typically difficult alloys to machine. Many attempts have been made to improve the machinability of these steels. These include the addition of additives such as sulphur, selenium and tellurium [22].


## 2.3  MILLING

### 2.3.1  General Features

The milling operation is one of the most important in the wide-ranging mechanical machining process [34], in the aeronautics, automotive and mould and die industries [13]. The fact that the milling cutter can take many different forms gives this operation versatility in terms of the geometries that can be generated [5]. This can be seen in table 2.8, which compares it with other machining processes.

Table 2.8: Comparison between milling and other machining processes for cutting flat, non-revolutionary surfaces [34].

| MILLING | APLAIMMENT |
| --- | --- |
| Cheaper operation | Cheaper machine and maintenance and shorter tool sharpening time |
| MILLING | EXTERNAL BROACHING |
| The broaching operation is impossible when the surface to be machined intersects with any other existing surface | Cheaper operation from a certain number of parts in the batch |
| MILLING | RECTIFICATION |
| Greater chip removal capacity | Improving the finish of the machined surface and achieving tighter tolerances. Often, the grinding operation comes after the milling operation. |

Milling is a process in which the removal of material from the workpiece is carried out intermittently by the rotating movement of a tool, usually a multi-cutting tool called a milling cutter. The tool moves at a feed rate in relation to the workpiece. A feature of the process is that each cutter edge removes a portion of material from the workpiece in the form of small individual chips [13].

Milling can be classified as concordant or discordant. This is seen when the cutting and feed movements have the same direction but opposite directions, and when the cutting and feed movements have the same direction and the same direction. This can be seen in figure 2.10, where the representation is **A** for concordant and **B** for discordant [32].

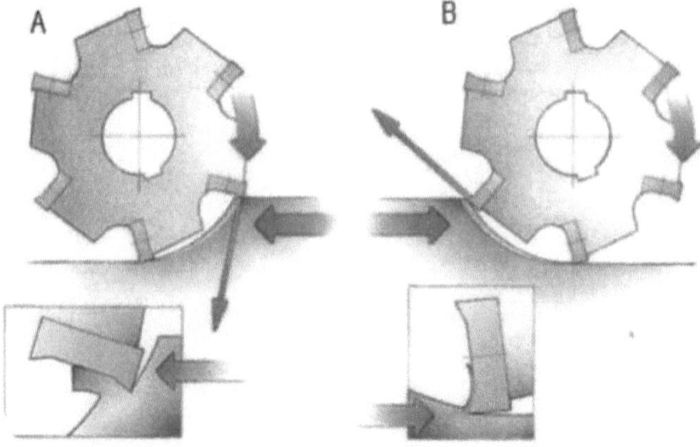


Figura 2.10: Cutting direction of peripheral milling [32].

In concordant milling, the cut starts by generating a thick chip, avoiding the effect of hardening due to plastic deformation. The large chip thickness generates a lower specific pressure and the cutting forces tend to push the workpiece against the tool, maintaining the cutting edge. In discordant milling, the chip thickness starts at zero, generating maximum specific pressure and high cutting forces, which tend to push the tool away from the workpiece. Initially, when the cutting edge touches the workpiece it is forced inwards - plastic deformation, creating excessive friction and high temperatures, causing the surface to harden [32].

According to Krabbe (2006), milling is a machining operation with chip formation that is characterised by [22]:

*A* multi-cutting tool, known as a milling cutter, has cutting edges which, in the vast majority of cases, are symmetrically arranged around its centre axis;

The milling cutter is equipped with a rotary movement and a feed movement to carry out the cutting operation, and each cutting edge (called a tooth) removes the amount of material that has been theoretically determined;

The macro-geometry of the workpiece is determined by the combination of the geometry of the cutter and the path of its movement over the workpiece.

Milling methods can be divided into two main groups: peripheral or tangential and frontal or flat. Other milling methods that exist can be considered variations of these two and depend on the type of workpiece and tool used [13].

In peripheral or tangential milling, the machined surface is generated by cutting edges on the periphery of the cutter and is generally a plane parallel to the tool axis. The cross-section of the milled surface corresponds to the contour of the cutter or combination of cutters used, described in figure 2.11 [13].

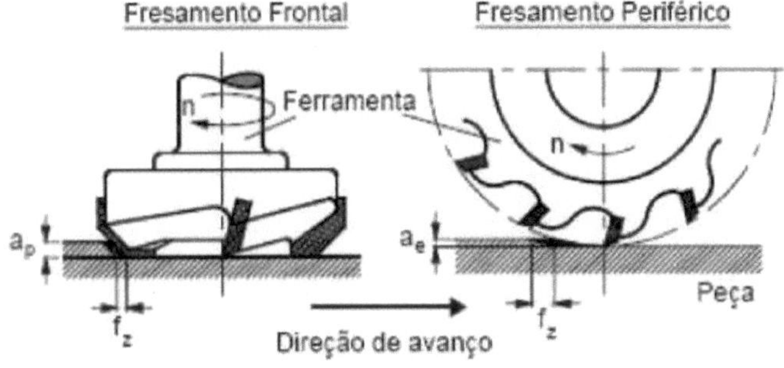


Figura 2.11: Front Milling and Peripheral Milling [13].

In face milling, figure 2.11, the machined surface is the result of the combined action of the cutting edges located on the periphery and the front face of the cutter, and is generally at right angles to the tool axis. Normally, in this type of milling, the milled surface is flat and does not correspond to the contour of the cutting edges [13].

In face milling (where the surface generated is perpendicular to the cutter axis) this definition of concordant and discordant milling cannot be fully applied. In symmetrical slotting (figure 2.12A) and ordinary face milling (figure 2.12B), the definition doesn't really apply, because in the first half of the contact between the cutter tooth and the workpiece, the cut thickness increases (which could be called a discordant cut) and in the second half of this contact, the thickness decreases (which could be called a concordant cut) [22].

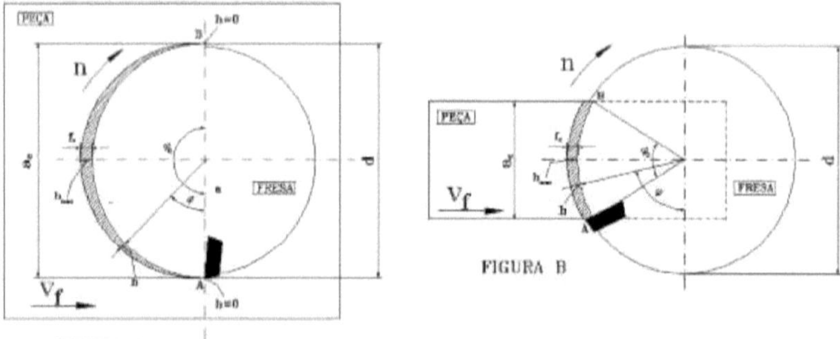


Figura 2.12:     (A and B): A. Symmetrical tear-face milling. B. Common Symmetrical Front Milling [22].


2.3.2 Movements between workpiece and cutting edge


In machining operations, all movements are important (figure 2.13). They are assigned directions, speeds and paths [34]. These are relative movements between the workpiece and the cutting edge. And they refer to the workpiece considered stationary - *instantaneous* movement, and two types of movement must be distinguished: those that directly cause the chip to come out - *effective cutting movement* - and those that do not take a direct part in chip formation [22].

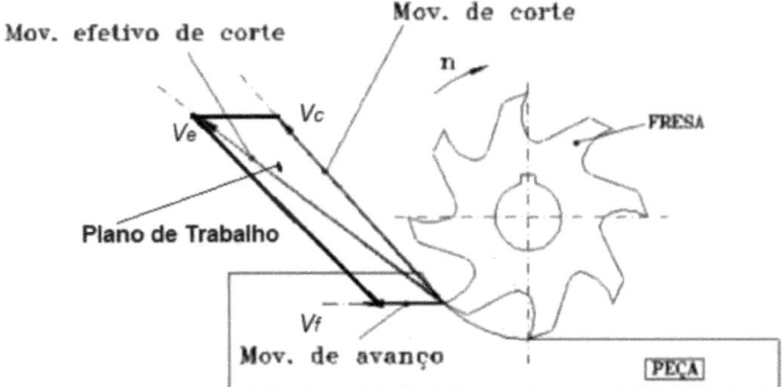


Figure 2.13: Instantaneous movement of the reference point at the moment of cut (adapted) [22].

### 2.3.3 Process quantities

In milling, the **cutting speed** $V_c$ is the instantaneous tangential speed resulting from the rotation n of the tool with diameter D at a point of contact with the workpiece, where the cutting and feed movements occur simultaneously [34]. A distinction must be made between **cutting speed** ($V_c$), **feed speed** ($V_f$) and **effective cutting speed** ($V^\wedge$) [22], figure 2.13.

Taking into account the following observations:

    *A* The cutting speed is the instantaneous speed of the reference point of the cutting edge, according to the direction and direction of the cut;

    *A* The feed rate is the instantaneous speed of the tool in the feed direction;

    The effective cutting speed is the instantaneous speed of the reference point of the cutting edge, according to the effective cutting direction.

The cutting speed can be obtained from the following formula:

$$V_c = \frac{\pi . D . n}{1000}$$

(2.1)

where:

$V_c$ = cutting speed (m/min);
**D** = cutter diameter (mm); **n** = tool speed (rpm).

The **feed** movement, $f$, consists of the movement between the workpiece and the tool, with the aid of the cutting movement, which produces continuous chip removal over several revolutions.

And the **feed rate** $V_f$ is the product of the feed rate and the rotation of the tool, which is given by:

$$V_f = f * n = f_z * Z * n;$$

(2.2)

where:

$V_f$ = mm/min;
$f$ = feed rate (mm/turn);
$n$ = tool speed (rpm);
$f_z$ = is the feed path per tooth per tool revolution measured in the feed direction (mm);
$Z$ = number of cutter teeth.

It is known that f is the **feed path** in each revolution of the tool, given in mm/turn [34]. So if $f_z$ is the feed path per tooth, we have:

$$f_z = \frac{f}{Z}$$

(2.3)

The **feed per tooth** corresponds to the distance between two consecutive machining surfaces, considered in the feed direction [34].

In order to deform a material during machining and achieve chip removal, the tool used must act with a certain force on the machined part. Knowledge of the magnitude and direction of the machining force, with its components in the cutting direction ($F_c$), in the feed direction (Fy) and in the direction of the tool axis ($F_p$), is of great importance in the design of machine tool elements [13].

The **cutting power ($P_c$)** is determined by the equation:

$$P_c = \frac{F_c * V_c}{60000} [KW]$$

(2.4)

where:

$f_c$ = cutting force (N);
$V_c$ cutting speed (m/min).

And the machining force ($F_u$), the force that acts on the cutting edge during machining, with its components $F_t$ active force, and $F_p$, passive force - which are perpendicular to each other. $F_u$ is divided into $F_c$, $F_t$, $F_{ap}$ (support force) and $F_e$, similar to $F_u$, generates the equation:

$$F_u = \sqrt{F_p^2 + F_t^2}$$

(2.5)

As in milling processes the feed direction angle tp is constantly varying, the resulting active force is expressed as the resultant of the components $F_{ap}$, $F_c$ and $F_t$ [34].

### 2.3.4 Productivity, Machining Rate.

The Machining Rate or material removal rate ($Q$ = cm³/min) is considered to measure productivity in terms of the amount of material removed by the machine tool in a specific period of time or specific volume of material removed [34,13].

In order to reduce machining times during wear, $Q$ is increased. This is achieved by increasing the axial depth of cut ($a_p$), the radial depth of cut ($a_e$) and the feed rate ($V_f$), which in turn depends on the feed per tooth ($f_z$), the cutting speed ($V_c$) and the speed ($n$). The relationship between the machining rate ($Q$) and the other cutting parameters is determined by [13,34]:

$$Q = \frac{a_e * a_p * V_f}{1000}$$

(2.6)

$V_f$ is described in equation 2.2.

The limits of the machine or type of application are restrictive factors for increasing the machining rate.

According to the Kienzle equation, as the feed per tooth ($f_z$) increases, the cutting force ($F_c$) increases exponentially. However, as the axial depth of cut ($a_p$) increases, the cutting force increases linearly. This equation is based on knowledge of the deformation stress and machinability of the materials, the machining cross-section and the number of teeth in the cut [13].

$$F_c = a_p . z_{ie} * k_{c1,1} * h_m^{1-mc}$$

(2.7)

where:

$z_{ie}$ = number of cutting edges;
  $h_m$ = average machining thickness (mm);
$k_{c1,1}$ = specific cutting force (N);
1-mc = Kienzle coefficient;
$a_p$. $z_{is}$ = is the sum of all the cutting edges acting simultaneously.

### 2.3.5 Preservation of the cutting tool

The parts to be machined can have the most varied shapes. This could complicate the machining process. However, thanks to the milling machine and its special tools and devices, it is possible to machine practically any part with surfaces of all types and shapes. A milling machine is a machine whose tool has a rotary movement and which allows the workpiece to be moved in one, two, three or more axes, linear or rotary. As such, it is a machine designed for the easy machining of prismatic parts - unlike the lathe, which mainly machines rotational parts with a profile of revolution [34].

Cutting speed ($v_c$) is the cutting parameter that most influences the life of a cutting tool, due to the friction characteristics between tool and workpiece. According to DE ABREU, Taylor demonstrated that the relationship between tool life (T) and

the cutting speed can be approximately expressed by the following equation [13,17]:

$$T = \left(\frac{C_t}{V_c}\right)^x$$

(2.8)

where:

T = tool life (min);
Ct = life for cutting speed of 1m/min (min);
x = exponent whose value depends mainly on the machine tool, tool and process.

As the feed per tooth (fz) increases, the tool travels along a shorter machining path and consequently rubs less against the workpiece material. However, as the machining thickness (hm) increases, tool life decreases beyond a certain value, determined by the loss of strength and consequent chipping of the cutting edge. Therefore, for the same machining rate, Q, lower cutting speeds and higher feeds per tooth are recommended [13].

Initial cutting edge/workpiece contact is considered unfavourable, depending on the region in which the first cutting penetration is established. In roughing, the axial depth of cut (ap) depends on the material being machined and the type of end tool [13].

## 2.4 CUTTING ECONOMIC CONDITIONS - VIEW ON COSTS

Nowadays, the global scenario has seen major competition between industrialised countries, i.e. for buyer markets. As a consequence, the profit of products is reduced to a minimum [14].

However, the quest to increase productivity in order to reduce production costs has become a key parameter in optimising the final cost of the product, especially when it comes to stainless steels, especially austenitic stainless steels [14].

These steels are of great interest to engineering due to their resistance to oxidation and corrosion, mechanical properties at high temperatures and toughness, as seen in section 2.1.4.

However, due to the plastic field and the high rate of hardening, these steels produce long chips and, when mechanically stressed, they harden forming a false cutting edge - PCA [14].

When a company has the strategy of producing faster, it has to give up savings on tool wear, so it can meet higher demands. If, however, your goal is to save on tool costs, demand falls and you rationalise the use of tools. In order to optimise the machining process, it is above all necessary to focus on productivity, tooling costs and machine hours. As input values change, calculations need to be updated to get the maximum return on the resources used. The choice of parameters used in the process must not ignore the strategy adopted by the company. This is why it is necessary to determine the balance between tooling costs and productivity [21].

For this reason, economic cutting conditions are currently one of the most investigated subjects within machining processes. The productivity of a machine tool or the total cost of manufacturing a part can be the criteria used to select cutting parameters (e.g. cutting speed) in order to improve the economics of a process [17].

### 2.4.1    Cutting Time

The machining cycle consists directly of the following stages in the manufacture of a single part [34]:

- placing and clamping the workpiece in the machine tool;
- approach and positioning of the tool;
- cut;
- removal of the tool;
- removal of the machined part.

In addition to these phases, they take part indirectly in the machining cycle (for a given batch of parts) [34]:

- Preparing the machine for machining;
- Removal of the tool for replacement or sharpening;
- Tool sharpening;
- Replacing the tool.

The total time of the machining process (Tt) of a workpiece:

$$t_t = t_c + t_1 + t_2$$

(2.9)

$t_c$ = cutting time;

$t_1$ = unproductive time (setting up, clamping and removing the workpiece, moving the tool closer and further away, preparing the machine);

$t_2$ = tool change and sharpening time.

the times of the three important parts of the process must be known:

$$t_t = t_c + t_s + t_a + \frac{t_p}{Z} + \frac{n_t}{Z} \cdot \left(t_{ft} + t_{fa}\right)$$

(2.10)

This equation is the total time to manufacture the part, adding up all the time spent in these phases, it looks like this, where [5,34]: $t_t$ = total time to manufacture the part $t_c$ = cutting time $t_s$ = secondary machining time $t_a$ = tool approach and removal time $t_p$ = machine set-up time $t_{ft}$ = tool change time

$t_{fa}$ = tool sharpening time

$Z$ =number of parts manufactured

$n_t$ =number of tool changes or sharpenings; formula: e:

$$n_t = Z \cdot \frac{t_c}{T} - 1$$

(2.11)

2.4.2   Optimisation of cutting conditions

When machining at low cutting speeds, the tool normally has a longer life, which means fewer tool changes and lower tool costs. However, machining time is high, increasing the cost of the machine tool and wages in relation to the volume machined. In recent years, the cost of tools and tool changes has increased to a lesser extent than the cost of wages and machine tools, and the reduction in costs is achieved through higher cutting speeds [17].

Once you have the equation for time, you can determine the equation for Cutting Speed for Maximum Production. However, it is necessary to simplify each part of this Y equation. By carrying out the necessary operations, [5] is obtained:

$$t_c = \frac{L_f}{V_f}$$

(2.12)

$$t_1 = t_s + t_a + \frac{t_p - t_{ft}}{Z}$$

(2.13)

$$t_2 = \frac{\pi \cdot d \cdot L_f \cdot V_c^{x-1}}{1000 \cdot f \cdot K}$$

(2.14)

And throwing it into the total time equation, where t1 is not linked to the cutting speed:

$$t_t = \frac{L_f \cdot \pi \cdot d}{1000 \cdot f \cdot V_c} + t_1 + \frac{\pi \cdot d \cdot L_f \cdot V_c^{x-1}}{1000 \cdot f \cdot K}$$

(2.15)

(2.15) Deriving as a function of speed, we will have the cutting speed for Maximum Production, as shown in figure 2.14:

$$V_{cmxp} = \sqrt[x]{\frac{K}{(x-1) \cdot t_{ft}}}$$

(2.16)

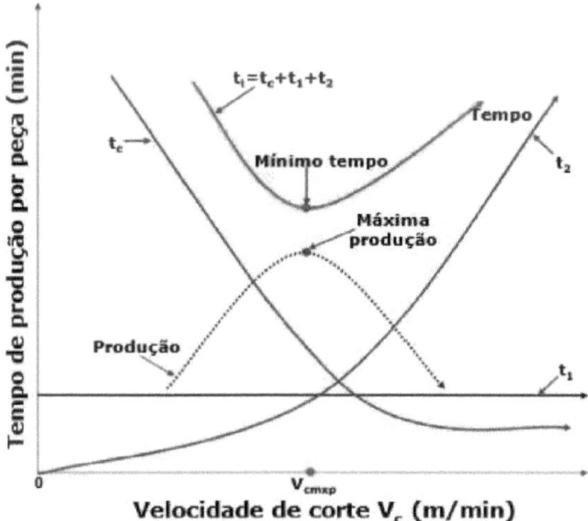



Figura 2.14:          Production time per piece as a function of cutting speed [24].

### 2.4.3 The Manufacturing Cost per Piece

The manufacturing cost per piece is made up of costs related to set-up and secondary times, which are treated as fixed costs, machine tool and operator costs, which are treated as the main cost, and the cost of the tools. As we can see in figure 2.15, as the cutting speed increases, the manufacturing cost per piece is predominantly influenced by the tool cost. This is due to the decrease in tool life, making the number of tool changes more frequent [17].

$$K_p = C_1 + K_{p1} + K_{p2}$$

(2.17)

Being:

*T T* Term independent of cutting speed

$$C_1 = \left(\frac{t_1}{60} - \frac{1}{Z}\right).C_2$$

(2.18)

Being:

*So s* Sum of labour and machine costs

$$C_2 = S_h + S_m$$

(2.19)

$K_{p1}$ = Cost of the machining process

$$K_{p1} = \frac{t_c}{60}.C_2 = \left(\frac{\pi.d.L_f}{60.1000.f.V_c}\right).C_2$$

(2.20)

$K_{p2}$ = Tool change cost

$$K_{p2} = \frac{t_c}{T}.C_3 = \left(\frac{\pi.d.L_f.V_c^{x-1}}{1000.f.K}\right).C_3$$

(2.21)

Being:

$C_3$ = Tool-related cost

$$C_3 = K_{ft} + \frac{t_{ft}}{60}.C_2$$

(2.22)

Being:

$K_{ft}$ = Tool cost per life

Substituting everything into Equation 2.32, you get:

$$K_p = C_1 + \left(\frac{\pi.d.L_f.C_2}{60\,000.f}\right).V_c^{-1} + \left(\frac{\pi.d.L_f}{1000.f}\right).\left(\frac{C_3}{K}\right).V_c^{x-1}$$

$$(2.23)$$

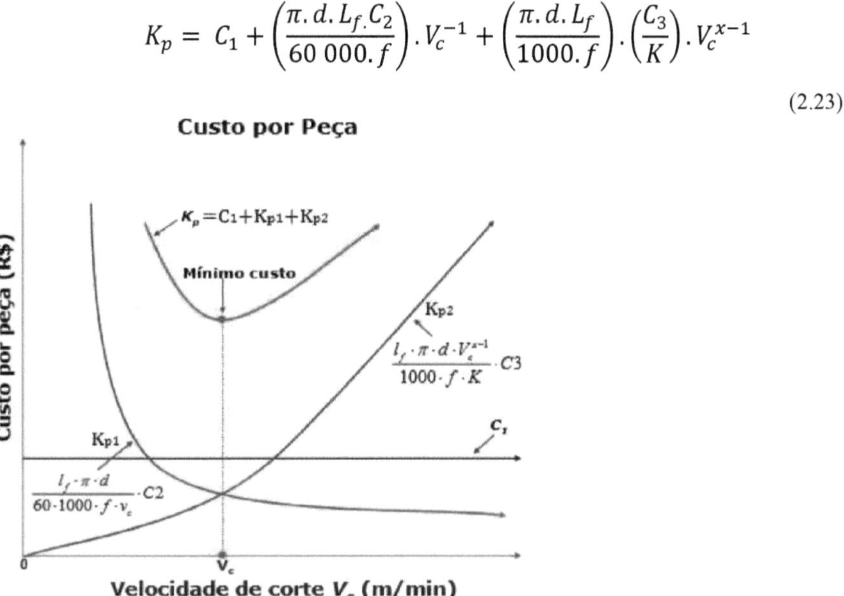



Figure 2.15: Cost per Part x Cutting Speed [24].

The minimum value of $K_p$ is obtained by deriving Equation 2.23 as a function of cutting speed and equalling zero (assuming $f$ and $a_p$ are constant) [16].

$$\frac{dK_p}{dV_c} = 0$$

$$(2.24)$$

$$-\left(\frac{\pi.d.L_f}{1000.f}\right).\frac{C_2}{60}.V_c^{-2} + (x+1).\left(\frac{\pi.d.L_f}{1000.f}\right).\left(\frac{C_3}{K}\right).V_c^{x-2} = 0$$

$$(2.25)$$

$$\frac{C_2}{60} = \left(\frac{(x-1).C_3}{K}\right).V_c^x$$

$$(2.26)$$

$$V_{co} = \sqrt[x]{\frac{C_2.K}{60.(x-1).C_3}}$$

$$(2.27)$$

And the forward speed is:

$$Vf = \frac{Vc.1000}{\pi.D}.n.fz$$

(2.28)

$$You.\,1000$$

### 2.4.4 Maximum Production Interval

The maximum efficiency range is defined as the interval between the Economic Cutting Speed and the Maximum Production Cutting Speed. It is very important that the cutting speed values used in machining processes are within this range, figure 2.15 [17].

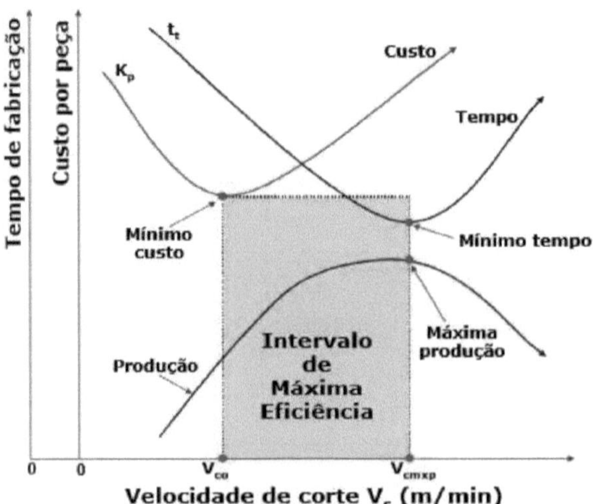



Figure 2.15: Representation of the Maximum Production Interval [24].

On a production line, in a period of high demand, the cutting speed that is closest to the optimum production time cutting speed should be chosen, but never higher, as the product delivery time is the most critical. In a period of low production, the cutting speed should be chosen as close as possible to the optimum cost cutting speed, but never lower than it [17].

# CHAPTER 3

MATERIALS AND METHODS

In order to have the data and confirm the proposed study, we began by calculating the speeds peripheral to the Maximum Production Interval - $V_{co}$ and $V_{cmxp}$. After this procedure, we fitted the curves and defined these limit points.

We can then determine the optimum cutting speed for production time and the optimum cutting speed for cost, taking into account the initial parameters: Machining Process (Milling and material used Superaustenitic Stainless Steel).

The purpose of this work is to present the concepts and development of the same for milling costs - *frontal and tangential*. The information presented here may vary in terms of composition, material values and actual market values.

- *Material of the part to be machined:* ASTM A744 CN3MN superaustenitic stainless steel. Chemical composition (%) - Table U composition limits: 0.74 Si; 0.015 C; 6.25 Mo; 20.79 Cr; 0.008 S; 24.65 Ni; 0.021 P; 0.63 Mn; 0.33 Cu; 0.2 N. $\sigma_r$ = 550 MPa; *Length* = 100 mm, *width* = 100 mm and *height* = 15 mm.
- *Cutting tool material:* Carbide Class K- WC+Co, KT490 LNHT 080408 PNR-IC 810, Rs = 0.8mm.
- *Operation:* $f$ = 60 m/min. Acceleration and deceleration 10 m/s$^2$ L = 200 mm. $fz$ = 0.54 mm/edge.
- *Cutting conditions:* Ø(cutter) = 150 mm. *Width* = 35 mm. Cutting edge = 24 cutting edges.

Cost data:
- *Operator salary:* 20.00 [R$/hour];
- *Machine hour:* 130.00 [R$/hour];
- *Cost per edge:* 14.50 [R$/aresta].

Additional Data:
- *Tool change time* tf = 1 min;
- *Secondary time* $_{ts}$ = 0.50 min/piece
- *Preparation time $t_p$* = **20** min
- *Lot size of parts to be manufactured* Z = 500 pieces

- *Approach and departure time* $t_a = 0.15$ min

Knowing the data presented and using the MATLab software programme, we obtain the equations for the Total Production Time and Total Production Cost curves. Then we have to derive these equations as a function of the cutting speed to find their minimum point, noting that this minimum point is within the limits of the Maximum Production Interval.

# CHAPTER 4

RESULTS AND DISCUSSION

The study material, CN-3MN superaustenitic steel, has a tensile strength equal to $\sigma_{rt}$ = 550 MPa. The cutting speeds (Economic and Maximum Production) are considerably higher than those found in SOUSA, 2011, p. 197 [5]. These speeds are totally related to the mechanical properties and composition of the material.

Economic cutting speeds and maximum production speeds could be tested for other materials and determined in the maximum efficiency range by comparing these two speeds. However, all this work has only been concerned with determining the maximum efficiency range for the steel in question, the superaustenitic CN-3MN.

The curves described below, in Figures 4.1 and 4.2, show the results only for the steel studied, with the aim of determining the Ideal Condition of the Production Process. Taking into account the choice that best suits the company, production radipez, $V_{(cxm)\,(p)}$, or manufacturing cost, $V_{co}$. It should be emphasised that comparisons can be made by studying other types of materials and obtaining their results.

The cutting speed can be observed in a range between 50 and 100 m/min and with a Maximum Production Speed $V_{cmxp}$ = 79.821 m/min, so we have the production time per piece tc = 0.805 min by equation 2.15, and we also obtain the Production Cost for this input value ($V_{(cmxp)}$= = 79.821 m/min) which is $Kp$ = 2.0449 R\$/piece in equation 2.23, as we can see in figure 4.1.

If we are aiming for high productivity, without worrying about the costs generated in the production of the material, it is advisable to use the maximum production speed, as the focus of this analysis is on the speed of the machining process.

As can be seen in figure 2.15, on a production line, in a period of high demand, you should choose the cutting speed that is closest to the optimum production time cutting speed, $V_{cmxp}$= , but never exceed it, because the product delivery time is the most critical.

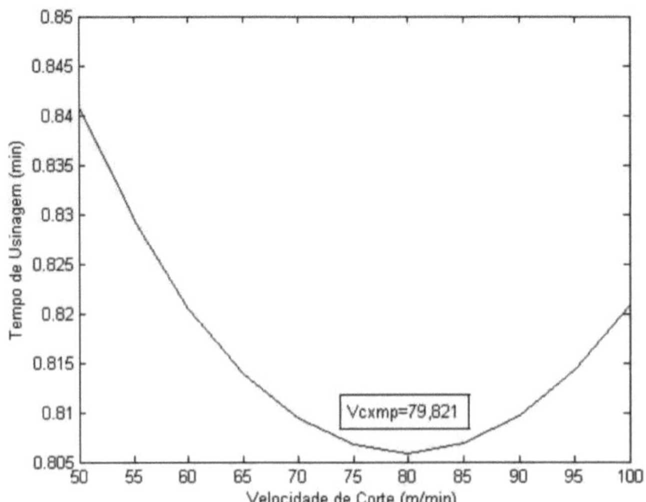

Figure 4.1 - Maximum Production Speed

And, the Economic Cutting Speed with a variation between 30 m/min and 80 m/min, as shown in figure 4.2, shows the Economic Cutting Speed, $V_{co}$= 54.40 m/min which implies a Production cost of Kp = 1.8477 R$/piece, generating a time cost of tc = 0.830 min per machined piece.

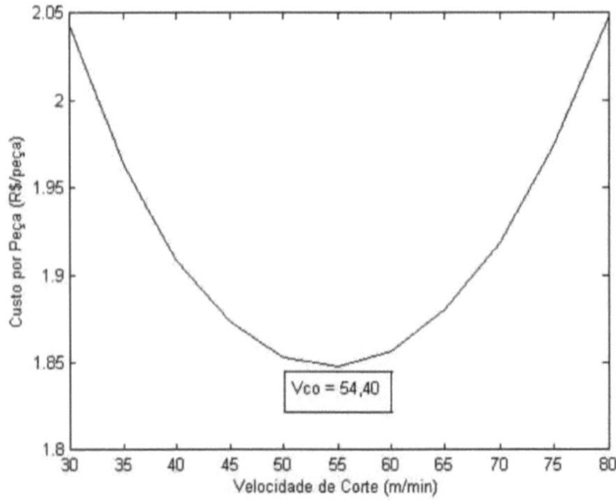

Figura 4.2  - Economical cutting speed

For the economical cutting speed, the production time shown was tt = 0.830 min, with the main parameter being the price of the material, even if production lasts longer.

44

And our Maximum Efficiency Interval graph is made up of the union of figure 4.1 and 4.2, shown in figure 4.3.

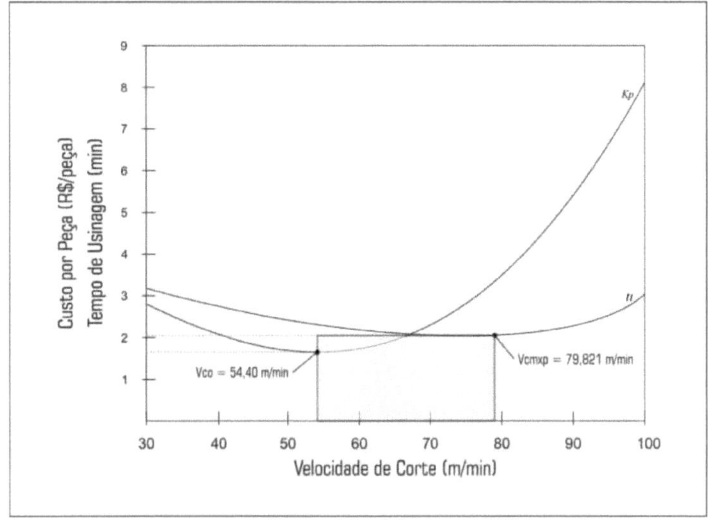


Figura 4.3  - Maximum Efficiency Range.

Comparing the two situations, if the speed is increased, productivity goes up as well; however, the cost of the part also goes up; the main factor being the life of the tool used. Conversely, by not using high cutting speed values, the integrity of the tool is better preserved, leading to fewer tool changes and production stoppages, and consequently the cost of the part decreases. However, the time it takes to make the part increases, which reduces productivity.

For those who don't want to decide only on high production or only on economy, the values in figure 4.13, in the Maximum Production Range, suggest values between these two extremes. Therefore, these values presented for the speeds of CN-3MN Superaustenitic Steel in the Milling process are of great help in obtaining an ideal situation in a given branch intention.

CONCLUSION

The results with regard to the economic aspects obtained, their evaluation and the machining parameters, were similar to those found in the literature, with those that determine Maximum Production Efficiency. And the introduction of the characteristic aspects of Austenitic Stainless Steels led us to a major problematisation. Due to the varying chemical composition and different mechanical properties that this material and its series possess. As a result, we had to choose a unique material, CN-3NM Superaustenitic Stainless Steel.

Cutting speed is an extremely important parameter in the construction of this scenario, as it allows you to choose the best route to take, with a specific profile for your production, despite the fact that it is restricted by design considerations and constants.

In the Maximum Production Range, these cutting speed values, chosen close to the lower limit of $7_{co}$, reduce the cost of production, because the cutting time is achieved, which increases the tool life. In other words, fewer production stops for tool changes and a smaller number of tools used mean that production costs fall.

On the other hand, by choosing speeds close to the $V_{cmxp}$ limit, a greater number of parts can be produced in a given time interval. Therefore, the Productivity parameter is increased. The consequence of this is a shorter tool life, as can be seen in figure 4.13, where costs start to rise again after a certain period of downtime, due to the period spent changing tools and spending unproductive time.

As mentioned above, the best way of ensuring production and not generating high costs is to be within the Maximum Production Range, not having high speed variations, but minimising them.

It is believed that this analysis opens up a range of studies for sectors that use CN-3NM Superaustenitic, companies that need a material with extremely high resistance to corrosion.

# CHAPTER 5

SUGGESTIONS FOR FUTURE WORK

- Analysis of the Machinability of Superaustenitic Stainless Steel CN- 3MN;

- Determination of the Specific Cutting Pressure for CN-3MN Superaustenitic Steel;

- Analysis of the Use of Cutting Fluid in the Machining of CN-3MN Superaustenitic Stainless Steel;

- Forms of Costing for Production with CN-3MN Superaustenitic Stainless Steel;

- Cost Considerations for Drilling Superaustenitic Stainless Steels.

# CHAPTER 6

BIBLIOGRAPHICAL REFERENCES

[1]  AB Sandvik Coromant, 2001. **Forget the claim that stainless steel is difficult to machine**. The World of Machining magazine, vol. 1.2001, p. 19-26.

[2]  AK Steel Corporation, 2004. AK Steel Comparator.

[3]  ASM Handbook - ASM International:
http://www.asminternational.org/documents/10192/1849770/06072G_Frontmatter.pdf

[4]  **BACIC, M.J.** *S* **Costa, E.A, (1995): "Costing System for a Small Parts Machining Company: A Practical Case", en: IV Congresso Internacional** de Custos, 16 to 20 October. Proceedings: Vol. 2, pp. 999-1018, Campinas, Brazil.

[5]  BARBOSA, Rômulo da Silva. **Cost Considerations in Ferritic Stainless Steel Turning**. 2015.

[6]  CARBÓ, Eng Héctor Mario. **Stainless steels: applications and specifications**. 2008.

[7]  CASAGRANDA, Marcio Veríssimo. **Study of the Machinability of AISI 303 Austenitic Stainless Steel**. 2004. 25f. Monograph (Mechanical Engineering Course Conclusion Paper) - Department of Mechanical Engineering, Federal University of Rio Grande do Sul, Porto Alegre, 2004.

[8]  Metal Mechanics Information Centre. **Classification of Structural Steels.**
**Codes            Codes**            Codes.                    Website:
http://www.cimm.com.br/portal/material_didatico/6337-classificaca/o-dos-acos-            structural-identification-codes#.WAZa8eArLIU. Accessed on 18/10/2016

[9]  CHIAVERINI, Vicente (2012). **"Steel and Cast Iron: General Characteristics, Heat Treatments and Main Types"**. São Paulo: ABM, $7^a$ Extended Edition.

[10]  CHILDS, Thomas. **Metal machining theory and applications**. 1st ed. New York: John Wiley & Sons Inc, 2000.

[11]  London     Metal     Exchange     (LME)     quotation.     Available     at:
http://www.maxiligas.com.br/cotacao-lme-london-metal-exchange#toplink.     Accessed     on     22 November 2016.

[12]  DA SILVA, Eduardo Miguel. **Study of the Correlation between Thermal Support, Magnetic Properties and Stress Corrosion in Welded Joints of AISI-409 Ferritic Stainless Steel**. Federal University of Itajubá. 2011.

[13]  DE ABREU, Pedro Jorge Moreira. **Analysis and optimisation of high-speed milling processes in the context of toolmaking**. 2010. 120f. Master's thesis in Solid Mechanics and Structures - Instituto Tecnológico de Aeronáutica, São José dos Campos.

[14]  DE CAMARGO, Robson. **Verification of the Machinability of Austenitic Stainless Steels through the Drilling Process**. Unicamp. 2008.

[15]  DE MAGALHÃES, Sávio Borba. **Economic Analysis of the Influence of the Cutting Fluid on the**

External Cylindrical Turning of 304L Steel. 2013.

[16]   DE SOUSA, Jhonatan Peres. **Study of the Machinability of Austenitic Stainless Steel.** 2014. UEMA.

[17]   EBERSBACH, Felipe Gustavo. **Time and Cost Optimisation of GG25 Grey Cast Iron Front Milling.** 2014.

[18]   FILHO, Fernando Teixeira. **The use of cutting fluid in the milling of 15-5PH stainless steel.** 2006. UNICAMP - FEM.

[19]   GENTIL V. **Corrosion.** 6. ed. Rio de Janeiro: Livros Tecnicos e Cientificos, 2011.

[20]   GRAVALOS, Márcio Tadeu. **Effects of machining on the surface integrity of a superaustenitic stainless steel** / Márcio Tadeu Gravalos. --Campinas, SP: [s.n.], 2008.

[21]   GUENZA , Jorge Eduardo. **Analysing the Milling Performance at High Cutting Speeds of GG25 Cast Iron in Industrial Applications.** 2008. UTFPR.

[22]   KRABBE, Daniel Fernando Moreira. **Optimisation of the Milling of Aeronautical Stainless Steel 15-5PH.** Unicamp. 2006.

[23]   MACHADO, Álisson Rocha. Da Silva, Marcio Bacci. **Metals machining** workbook. 1999.

[24]   MENDES, Ronã Rinston Amaury. **Study on Minimising the Cost of Machining Hard Steel Using Response Surface Methodology.** Federal University of Itajubá. 2006.

[25]   NÚCLEO INOX. Collection of Technical Information - Stainless Steel. Website: http://www.nucleoinox.org.br. Accessed on 21/10/2016.

[26]   OLIVEIRA, Leonardo Albergaria. **Influence of the Addition Metal on the Susceptibility to Stress Corrosion of Dissimilar Welded Joints of 316 Austenitic Stainless Steel and 2304 Duplex Stainless Steel.** Itajubá, 2013.

[27]   PADILHA, A. F., GUEDES, L. C, **Austenitic Stainless Steels: Microstructure and Properties.** Editora Hemus. 1994.

[28]   PEREIRA, Janaína Aparecida; FERREIRA, Camila Corrêa Martins; DA SILVA, Márcio Bacci. **Analysing the Turning of ABNT 304 Stainless Steel through Chip Temperature.** 2009

[29]   Portal Metálica. Construction. **Structural Steel.** Website: http://wwwo.metalica.com.br/acos-estruturais. Accessed on 18/10/2016.

[30]   RAMOS, Julio Endress. **Stability characterisation of austenitic stainless steels without added nickel.** 2009.

[31] RITONI, Marcio. MEI, Paulo Roberto. MARTINS, Marcelo. INOX: Physical Metallurgy. **Effect of ageing heat treatment on the microstructure and impact properties of ASTM A 744 Gr. CN3MN superaustenitic stainless steel.** 2010.

[32] ROHLOFF, Ronaldo Carlos. **Effect of Cutting Parameters on the Milling of AISI 420 Stainless Steel for Moulds and Dies.** 2012.

[33] SEDRIKS, A. J. **Corrosion of stainless steel.** John Wiley Sons Inc, USA, n. 2, p. 47-53, 1996.

[34] SOUZA, André João de. **Machining Manufacturing Processes** - Part 2. UFRS, 2011.

[35] SOUZA, Antônio Carlos de; NOVASKI, Olívio; OLIVEIRA PAMPLONA, Edson de; BATOCCHIO, Antônio. **Economic Conditions in the Machining Process: An Approach to Considering Costs**. State University of Campinas and Federal School of Engineering. Campinas-SP. Website: http://www.intercostos.org/documentos/Trabajo100.pdf, Accessed on 24/11/2015.

[36] Federal University of Paraná. 2010. **Metallic Structures**.

[37] VALERIANO, L. C. **Influência da Precipitação de Fases Secundárias na Resistência à Corrosão do Aço Inoxidável Super Duplex uns S32520**. Federal University of Itajubá. p. 30-35. 2012.

[38] **Wu, S.M.; Ermer, D.S. (1966): "Maximum Profit as the Criterion in the Determination of the Optimum Cutting Conditions". Journal of Engineering for** Industry, Transactions of the ASME, Nov., pp.435-442.

## ANNEX 01 - CALCULATION MEMORIAL

Using the data from item 3 - Materials and Methods and the equations from the Benchmark Theoretical in **2.4**, below is the Calculation Memorial.

*Advance, Depth and $C_v$*

According to equation (2.2):

$$f * n = fz * z * n$$

$$f.n = 0,54.24.n$$

$$f = 12,96 \ mm/volta$$

$$a_p = 5 \ mm$$

$$C_v = 213,75$$

*Tool life*

$$T = \left[\frac{60^y}{V_c} \cdot \frac{C_v}{(a_p.f)^i} \cdot \left(\frac{a_p}{5f}\right)^g\right]^x$$

$$T = \left[\frac{60^{0,2}}{V_c} \cdot \frac{213,75}{(5.12,96)^{0,28}} \cdot \left(\frac{5}{5.12,96}\right)^{0,14}\right]^5$$

$$T = \left[\frac{105,325}{V_c}\right]^5$$

*Total Production Time*

$$t_t = \left(\frac{\pi.d.L_f}{1000.f.V_c}\right) + t_1 + \left(\frac{\pi.d.L_f}{1000.f.V_c}\right).\left(\frac{t_{ft}}{T}\right)$$

$$t_t = \left(\frac{\pi.150.200}{1000.12,96.V_c}\right) + \left(0,50 + 0,15 + \frac{20 + 1,0}{500}\right) + \left(\frac{\pi.150.200}{1000.12,96.V_c}\right).1,0.\left(\frac{V_c}{105,325}\right)^5$$

$$t_t = 0,692 + 7,2722.V_c^{-1} + \left(\frac{7,2722}{105,325^5}\right).V_c^4$$

Maximum Production Cutting Speed

$$V_{cmxp} = \sqrt[x]{\frac{K}{(x-1).t_{ft}}}$$

$$V_{cmxp} = \sqrt[5]{\frac{105,325^5}{(5-1).1}}$$

$$V_{cmxp} = 79,821 \text{ m/min}$$

Total Cost of Production

$$C_1 = \left(\frac{t_1}{60} - \frac{1}{Z}\right).(S_h + S_m)$$

$$C_1 = \left(\frac{0,692}{60} - \frac{1}{500}\right).150$$

$$C_1 = 1,43$$

$$K_{p1} = \left(\frac{\pi.d.L_f}{60.1000.f.V_c}\right).C_2$$

$$K_{p1} = \left(\frac{\pi.150.200}{60.1000.12,96.V_c}\right).150$$

$$K_{p1} = \frac{18,18}{V_c}$$

$$K_{p2} = \left(\frac{\pi.d.L_f}{1000.f}\right).\left(\frac{C_3}{K}\right).V_c^{x-1}$$

$$K_{p2} = \left(\frac{\pi.150.200}{1000.12,96}\right).\left(\frac{17}{105,325^5}\right).V_c^4$$

$$K_{p2} = \left(\frac{123,63}{105,325^5}\right).V_c^4$$

$$K_p = C_1 + K_{p1} + K_{p2}$$

$$K_p = 1,43 + \left(\frac{18,18}{V_c}\right) + \left[123,63.\left(\frac{V_c^4}{105,325^5}\right)\right]$$

Economical cutting speed

$$V_{co} = \sqrt[x]{\frac{C_2.K}{60.(x-1).C_3}}$$

$$V_{co} = \sqrt[5]{\frac{150.105,325^5}{60.(5-1).17}}$$

$$V_{co} = 54,40 \text{ m/min}$$

## yes I want morebooks!

Buy your books fast and straightforward online - at one of world's fastest growing online book stores! Environmentally sound due to Print-on-Demand technologies.

Buy your books online at
**www.morebooks.shop**

Kaufen Sie Ihre Bücher schnell und unkompliziert online – auf einer der am schnellsten wachsenden Buchhandelsplattformen weltweit! Dank Print-On-Demand umwelt- und ressourcenschonend produziert.

Bücher schneller online kaufen
**www.morebooks.shop**